Electric

The Ultimate Guide for Understanding the Electric Car and What You Need to Know

Table Of Contents

Introduction

This short book is for people who are interested in learning more about electric cars and are not sure where to start or what information to rely on. It was published in response to the high demand of people wanting to know more about electric cars and why they are becoming so popular. The internet has a ton of articles and misinformation about electric cars that confuse people who are interested in learning about this revolutionary craze and possibly interested in investing in an electronic vehicle for themselves.

This book traces the recent developments, innovations, advantages and disadvantages, as well as the future of the electric car industry. By understanding this information, you will see why the costs for commercial electric cars are what they are. We will also go into a short overview of current costs for electric cars for those who are interested.

It is recommended that you take notes while you are reading this book. This will ensure that you get the most out of the information in here. The notes will help you to pinpoint exactly what you need to remember and by writing things down, you will be able to recall specifics if you are looking to purchase.

Lastly, remember that it is encouraged that you do your own research on the topics you want to look deeper into. The more you understand about electric cars and what the pros and cons are, the more educated your decision-making process will be.

Chapter 1:

The History of Electric Cars

Unbeknownst to many, electric cars have been around for quite a while now. Although it would seem that electric cars are recent innovations, they are not. In fact, these brilliant inventions have been around for over a hundred years.

Models and Prototypes (1801-1850)

The earliest models of electric-powered vehicles were built in Scotland. It was the Scottish Robert Anderson who first introduced the electric-powered carriage. Anderson experimented on different prototypes from 1832 to 1839 before he was able to build a working model.

In 1834, a group of scientists led by Thomas Davenport from the United States invented a car model that operated using direct current electric-powered motor. The vehicle was designed to run through an electrified circular field.

The First Age (1851-1900)

It was in the late 1800s when England and France first developed electric cars, while the American public's attention was not captured by electric cars until 1895. Not to be outdone, Germany in 1888 build the world's first four-wheeled electric vehicle. This endeavor was initiated by Andreas Flocken, an engineer and a physicist.

This was followed by several innovations, as interest in motor vehicles significantly increased during the latter part of 1890 and the early part of the 1900s. In 1897, the United States spearheaded the very first electric cars' commercial application with a fleet of taxis in New York City. In the same year, Pope Manufacturing Co. established its name as the first wide-scale electric vehicle maker – including the taxi fleet of New York City, among others.

After decades of continuous development, France finally joined the fray and released the "La Jamais Contente", which travelled continuously at a speed of more than 100 km per hour – a milestone for the electric automotive industry. By the 1900's, the United States fully embraced electricity-powered cars. In fact, at least 28% of the vehicles sold that year were powered by electricity.

The Peak and the Trough (1901-1950)

Early models of electric cars can be likened to horseless surreys and carriages that were electrified. Wood's Phaeton of 1902 was one of these early models. This early electric vehicle had a maximum speed of 14 miles/hour with an 18-mile range. Wood's Phaeton retailed at approximately $2,000 during that time.

Later on, motor vehicles became more popular as they were set available in electric, gasoline, and steam. Despite having other alternatives aside from electric power, in 1899 and 1990, electric powered vehicles outsold other car types in the United States.

This was primarily because electric powered cars held several advantages over the rest of the car types during this time period. Steam powered cars, for instance, had very long start-up time, which could reach up to 45 minutes, especially lengthy during cold days. Moreover, electric powered cars had longer range than steamed powered vehicles at this time. This was due to the latter's need to refill water often.

Electric vehicles were also seen as far superior to gasoline based-cars during this time. Electric powered cars did not have noise, vibration, and/or smell that were usually experienced with gasoline powered cars. In addition, gear shifting in gasoline powered cars made them difficult to drive. In a nutshell, electric cars were more popular than all other car types, largely because manual start-up efforts were not necessary and there was no need to struggle with shifting gears. Add to this the fact that there were few good roads during this time; hence, the short range of local commuting, made electric cars' reach just right.

During this time, a person would need $1,000 to be able to avail the basic electric car. Nonetheless, most electric cars produced during this time were massive and ornate carriages, designed by and for the upper class of society. Such elaborate cars were made from expensive materials and the interiors were fancifully ornamented.

In 1910, a person would need at least $3,000 to purchase an electric powered car. This could perhaps be attributed to William Howard Taft becoming the first President of the United States to have bought his own electric vehicle in 1909. This led to the steep rise of popularity of electric vehicles in the country. Taft purchased a Baker Electric, one of the best models at that time.

The popularity and success of electric vehicles were at its peak until the 1920s. However, time changes everything. In fact, a year before Taft bought his electric vehicle, the car market saw the first glimpse of petrol-powered car: Ford Model T. This car model, the first of its kind, would eventually cause the downfall of electric vehicles.

There were social changes that pushed the success of electric vehicles aside. One of these factors was the arrival of better road systems in the United States, around the 1920s. The new road systems made the connection between big cities; hence, traveling now required vehicles capable for long-range travels.

The problem with hand cranks was also eliminated by the electric starter invention. By 1912, global sales of electric vehicles peaked at 30,000 cars which would have spelled good news to electric car manufacturers had they not been faced with the threat of petrol-powered cars.

To add to electric vehicle makers' already growing woes, the price of gasoline then became affordable, even to average consumers, because of the Texas crude oil discovery. Also, the mass production of vehicles with internal combustion engines made such cars widely available and reasonably priced. Henry Ford was the one behind the said initiative, to mass produce the cars, which were priced at $500 to $1,000. Meanwhile, the price of electric cars continued to rise; making it less preferable for the average consumer.

It was in 1935 when electric car production and development slowly started to disappear. Through innovations and discoveries, the disadvantages of gasoline fueled cars were slowly eliminated. Consequently, the electric car industry died until the 1960s.

Largely due to war and oil rationing in Japan, leading Japanese carmaker Tama was forced to release an electric car powered by 40V lead acid battery.

The Rebirth of the Electric Vehicle (1951-2000)

The gasoline powered cars, however, increased the United States' dependence for foreign crude oil. Moreover, problems have been encountered with the internal combustion engine's exhaust emission. Such issues in the 1960s and 1970s brought about the need to seek for alternatively fueled vehicles.

Six years after the death of the electric car industry, the U.S. Congress passed a legislation that proposed electric vehicles as a way to curb the worsening air pollution in the country. Hence, from that time up to the present day, there have been varying attempts to produce practical electric cars.

In 1964, the first electric truck with a range of 62 miles and a speed of 25 mph was developed by

Battronic Truck Company. The said company was formed when Boyetown Auto Body Works, Electric Battery Company's Exide Division and Smith Delivery Vehicles Lmt., joined forces. The Battronic electric truck could carry up to 2,500 pounds of cargo.

In 1973, electric vehicles enjoyed renewed interests due to long queue at oil-filling stations and the constant rise of oil prices, mainly because of the OPEC oil embargo. From that year, up to at least 10 years after, government agencies instituted further efforts to showcase the competence of battery powered vehicles. Battronic teamed with General Electronics during that time to produce 175 vans to be used in the utility industry. In addition to that, during the 1970s, Battronic created 20 public buses.

It was not long before other companies started developing and producing electric powered cars as well. For instance, Sebring-Vanguard, one of the leading companies in the electric car industry, produced more than 2,000 electric cars called "CitiCars." The range of the car was 50 to 60 miles and the top speed was from 38 mph to 44 mph. Another leading electric car company, Elcar Corporation, produced "Elcar" which reached speeds of 45 mph and a range of 60 miles.

The re-emerging electric car industry started to become noticeable, not just to the public, but also to the United States Government. The Postal Service of the United States bought 350 electric powered delivery jeeps in 1975. The American Motor Company was then the manufacturer of the cars. The electric delivery jeeps had a 40-mile range and reached speeds of 40 to 50 mph. These vehicles were used in the test program of the United States Postal Service.

In 1976, the French government initiated research and development focused on electric vehicles by launching a program called "PREDIT". They gave scholarship grants to any French academician who was willing to help develop electric cars.

Aside from the growing number of electric cars that were being manufactured, regulatory and legislative actions during the 1990s also served as elaborate efforts to the further developments of electric vehicles. Some of the enacted laws included the Clean Air Act Amendment of 1990 and the Energy Policy Act of 1992. The regulations laid strict requirements for air emissions and pursued the reduction of the gasoline use. In addition to this, several states across the United States issued the requirements of zero emissions for vehicles.

It was not long until the Department of Energy, together with the leading electric car and conversion companies, established the Partnership for a New Generation of Vehicles. PNGV was aimed at pursuing active efforts of developing electric vehicles. This paved the way for the development of electric cars and converted electric cars that could reach the speeds of 50 to 150 mph between charging. The Chevrolet S-10 truck was an example of the said cars. The S-10 could reach 60 miles for its range and could be recharged in less than 7 hours.

Several other car models were released in response to the growing electric car industry. Solectria's Geo Metro was one of these models. Geo Metro was a converted, 4-seater, electric car. During the American Tour de Sol of 1994, Geo Metro traveled over 200 miles without recharging. The car used ovonic nickel metal hydrate batteries to operate.

In 1996, General Motors, one of America's largest car manufacturers, mass-produced the EV1, the company's first electric-powered vehicle. This was produced to comply with the 1990 legislation in California, requiring car manufacturers to produce a Zero Emission Vehicle (ZEV). In the following year, Japanese car maker Toyota introduced the world's first commercially-available hybrid vehicle: the Prius. At least 18,000 units were sold in 1997 alone.

Ford developed electric vehicles which could reach speeds of up to 75 mph. One of which was the Ford Ranger electric version, with acceleration from 0-50 within 12 seconds and could carry up to 700 pounds of payload. Unlike Ford, General Motors decided to produce electric cars from scratch instead of converting their existing models. GM EV1 became one of their creations. This electric car could reach a top speed of 80 mph. It could also accelerate from 0-50 mph in less than 7 seconds.

Several other electric car models became available, particularly in 1998. Recent models are now powered with advanced battery packs, like the nickel metal hydride and lithium-ion. Indeed, the developments in the electric car industry make their products more competitive when it comes to satisfying driving requirements. Nonetheless, until the publication of this book, electric car prices range from $20,000 to $40,000 on the low end, making them less affordable for the average consumer. The price, however, can be lowered significantly when incentives and tax credits are counted in.

The Millenium (2001 – present)

Government and private entities began working hand-in-hand to reintroduce electrification of vehicles. In 2008, thanks to the political and religious unrest in many Middle Eastern countries, oil prices hiked to more than $145 per barrel. This outrageous price led to a steep rise in the prices of commercially-available oil all over the world.

Two years later, Nissan responded to high gas prices by introducing the Nissan LEAF. In 2011, the French electric car sharing service and the largest of its kind in the world, Autolib launched its service in Paris, unveiling a stock of a staggering 3,000 electricity-powered vehicles loaned to interested Parisians.

The same year saw the global stock of electric vehicles rise to a new peak of 50,000 units sold the world over. The government consortium in France promised to buy 50,000 electric vehicles over the next four years. Nissan LEAF won the European Car of the Year award a year after it was first introduced.

In 2012, General Motor's Chevrolet Volt climbed the US market sales charts from virtually unknown to outselling at least half the car models sold in the country. This saw the rise of global stock of electric vehicles to 180,000 units.

In 2014, for the first time, a Euro NCAP 5-star safety rating car model that is autopilot-ready and with an available multiple-wheel drive double motor, which allows a range to a maximum of 330 miles, was introduced. The Tesla Model S is hailed as one of the best electric-powered cars to be produced.

Unfortunately, several car manufacturers in 2015 were caught cheating on their emission regulations claim. The scandal has led more people to believe that the single best way to decrease gas consumption and smoke emissions is through electric vehicles. At this point the number of electric vehicles sold around the world reached 700,000. In the United Kingdom, 22,000 vehicles were sold to the USA's 275,000.

Chapter 2:

Under the Hood

An electric vehicle, just like other car types, has a motor. However, in an electric vehicle, the motor is powered by battery packs that are rechargeable instead of a gasoline engine.

The physical appearance of an electric car from the outside does not make it distinguishable from a gasoline powered car. Some electric cars are even converted from a previously gasoline powered car. Most of the time, the only distinguishable clue observable to the untrained eye, is the silent nature of the electric vehicle compared to its counterpart.

In total, there are three main parts working together under the hood of an electric car. These parts are the following; electric motor, controller, and a rechargeable battery. Electric powered cars operate on the principle of electricity/current. The rechargeable battery provides power to the controller and the controller gives power to the motor. Consequently, the power received by the motor from the battery will then be sent to the transmission, which will cause the wheels to turn.

Parts of an Electric Car

Potentiometer

Hooked to an accelerator pedal is the *potentiometer*, also known as the variable resistor. This is the part responsible for providing the signal that dictates to the controller the amount of power that is to be delivered.

There are two potentiometers installed in an electric car. This is for safety purposes. When the two potentiometers send in different signals, the controller will not react.

Battery

The power of an electric car comes from the *battery*. This part is the one responsible for sending power to the car's controller.

The most commonly used battery packs are nickel metal hydride, lead-acid, and lithium-ion batteries.

Controller

The *controller*, after receiving the power from the battery, will send that power to the motor. When the vehicle is stopped, the controller is sending zero power. Consequently, the controller sends full power when the driver hits the accelerator pedal.

Aside from zero power and full power, the controller can also deliver powers in between the two levels. The power regulation of the controller will depend upon the potentiometer.

Motor

The power sent by the controller will be received by the *motor*. This part is responsible for activating the transmission, which will then cause the wheels to turn.

Safety Guidelines

Working on electric cars is similar to working on their petrol-powered counterparts. In both situations, safe working practices must always be followed. As in the case of working on or maintaining high-voltage systems, extreme caution must be practiced.

People who know little about the system must let the professionals handle the job; otherwise, they risk contracting injury or worse, death. Batteries and motors of electric vehicles contain high magnetic and electrical potential that, when handled carelessly or incorrectly, can cause irreversible injury.

Often as a precaution, the design of electric cars use the color red or orange for high-voltage wires. At all times, electric car owners must follow the instructions prepared by the car manufacturer.

Before Maintenance:

The ignition switch must be turned OFF and the key should be removed.

The electric system must be drained of power by turning the battery module switch OFF.

Wait for the storage capacitors to be fully-discharged before carrying out maintenance operations on the electric system. This usually takes at least 5 minutes.

During Maintenance:

Insulating gloves long enough to cover the hands and half the length of the hands and elbow must be worn at all times.

Likewise, only properly insulated tools must be used when repairing the system to significantly cut down the risk from being electrocuted and minimize exposure from high-voltage systems.

The place where maintenance operations are performed must be well-lit to prevent incorrect re-wirings which may lead to accidental (and fatal) short-circuits.

Interruptions/Breaks:

Before leaving a disassembled motor or any exposed high-voltage equipment, it should be made sure that:

The switch connected to the battery is turned off.

The key is removed and the ignition is likewise turned off.

The work area is inaccessible to persons who know nothing about the electric system.

The work area must be kept lit to prevent any accidental touching of exposed electric parts.

After Maintenance:

All repairs have been checked and completed.

High-voltage terminals and their corresponding wirings are not shorted or damaged to prevent any untoward incident involving the system.

Terminals and connections are adequately tightened to their corresponding torque.

The reassembled parts of the vehicle's body should be carefully insulated from high-voltage terminals. This must be double-checked before turning the battery module on.

Risk Management

Risk: Electric Shock

Preventive Practice: When working on a high-voltage system, mechanics must always use insulated gloves and insulated tools. Power tools must also be in pristine condition. Also, only mechanics who have ample experience and training handling high-voltage systems should work on electric vehicles.

Risk: Battery Acid Leak

Preventive Practice: Personal protective equipment must be kept in good working condition, especially when dealing with highly-corrosive and toxic substance like sulfuric acid. Mechanics who are working on batteries must, at the minimum, wear rubber gloves and rubber apron; however, it is ideal to wear overalls and goggles as well.

Risk: Short Circuits

Preventive Practice: When dealing with high voltage connection, it is best to use a jump lead (with in-line fuse). This hinders the possibility of a short during connection testing. Immediately disconnect the system from the battery should the possibility of a short circuit exist.

Risk: Starting Engines

Preventive Practice: If overalls are not available, mechanics must wear clothes tight enough to allow them to move freely. The keys must always be kept inside the pockets to avert any instances where someone else starts the engine while the mechanic is working on the engine.

Risk: Lifting Vehicles

Preventive Practice: Car wheels must be chocked or brakes must be applied before raising the vehicles using a jack lift.

Chapter 3:

Development and Innovations

The internal combustion engine cars remain the most popular car utilized all over the world. However, with the continuous increase in oil prices and the growing concern about the environmental degradation, heightened by the use of fuel to power cars, more people are considering alternatives than ever before. Electric vehicles are the one promising alternative for the conventional car type. Hence, efforts concentrating on improvements to the car's capabilities are already mounting.

Currently, several developments are already being seen when it comes to the car's design. For instance, the electro motor of an electric car can be situated in the front or back of the car. Moreover, electric vehicles can also be powered directly in its wheels. This is made possible by small electric motors positioned in the car's wheels.

The power, as mentioned earlier, comes from the battery, which can be located at the back, middle, or front part of the car. Design is one of the areas that innovative electric car companies are aiming at; hence, it can be expected that these companies will be releasing many more models with upgrades on design in the next decade.

The electric vehicle's battery is also very important; thus, major innovations cover the improvements related to it. Currently, Li-ion polymer batteries are being used to power electric vehicles. This battery showed significant advantage in power density over the conventional batteries used, like Ni-MH and Pb-Ac. Research, however, hasn't stopped in creating better batteries for the electric cars.

Giant car companies like Toyota, Nissan, and Renault have teamed up with other companies in order to create electric vehicle batteries with cycle-life, able to resist fast-charging, increased energy density and improved power output. Innovations have already produced cost efficient, durable, and lighter batteries; nonetheless, they actually are still relatively heavy and expensive to produce. Car companies are aiming at smaller battery packs to avoid vehicle weight and expensive vehicle costs.

Another area that can be improved is the transmission unit. Most electric vehicles available today use a central electro motor that sends power indirectly to the car's wheels. Innovations are now being conducted to improve electro motors by increasing their power and decreasing their size and weight.

In addition to all of this, the electric vehicles' energy infrastructure also needs improvements. Once the use of electric vehicles will be implemented in large scales, the need for more advanced and improved charging infrastructure will be increased. Innovations in this particular area are aimed at the possibility of charging electric vehicles in parking lots in a shorter period of time.

In relation to this, systems for automatic payment must be set available in parking spaces. Currently, the main infrastructure required for electric vehicles' power supply is not widely available. If an electric car's battery is empty, it could be plugged into any available power point to recharge, which would drastically increase convenience.

An electric car fully charged from the standard 230-volts power point can reach the 200-300 km range, once fully charged. Some electric vehicles that are used for short travels can be plugged into the driver's household during the night time in order to be utilized in the following morning.

Of course, the power output and battery size will influence the charging time. Usually an electric vehicle needs to be charged for 2-7 hours. Naturally, another innovation that is being considered is having stations where batteries can be traded, which will run automatically.

Battery of the Future

With over 60 partners from both industrial and academic circles including Volkswagen, Evonik, BASF, Li-Tech, and Bosch, the Lithium-Ion Battery Alliance was founded in 2008 to develop cost-effective, high-capacity lithium-ion batteries for automobiles. Among the more successful initiatives are the following:

HE-Lion Project

The team – the largest in the alliance – works on creating electric car batteries that have a high energy density level. Among the other goals they are hoping to meet is to build a battery that is environment-friendly, long-lasting, and ensures a good level of safety.

Li-Redox Project

Battery safety is considered to be the most apparent concern for many car manufacturers, and this is the mandate being carried out by researchers in this project. Creating a battery that can store high quantities of energy is not without considerable risk. Risks include possible charge instability and temperature fluctuations.

KoLiWin Project

Researchers working on this project are looking for and testing new varieties of gel polymer electrolytes that can bring good mechanical strength to the battery aside from what could be high conductivity levels.

Electric-Powered Race Car Motor

The image of electric-powered hybrid cars constitute a slow-moving car that would not hold water when raced against petrol-run cars. That's about to change as the engineers at the University of Oxford have successfully developed electric motors that are more than capable of running at par with many of the world's fastest sports cars.

Creating an electric motor that can match the capabilities of petrol-powered motors seemed unthinkable, but Oxford student Tim Woolmer and his professor Dr. Malcolm McCulloch were both challenged to make possible what seemed to be impossible.

Electric motors have two main parts: one that's stationary and one that's always moving. Electric currents and magnetic fields cause the moving part to spin but the geometric configuration of the rotating drum leads to high speed losses. The team led by Woolner and McCulloch found a system that makes the rotating part move similar to a circular pancake.

McCullock devised a way to create the complicated shapes required to turn their system into a reality. After building the first prototype, the benefits were apparent. The new motor decreased electric losses by as much as 25 percent, which then reduced the required cooling by just as much.

Not yet satisfied with what they managed to build, the team always found a way to improve the system to further better its efficiency and thus its application. After several years of painstaking research and development, the Oxford team managed to create segmented armature motor that is several times lighter than its nearest competitor.

The motor became so popular that Oxford Yasa Motors, a company comprised mainly by the members of the Oxford team, received a 3.4 million-euro grant in 2009 for them to develop the design into a real product for the market. Delta Motors has since been using the motors they designed.

Google Self-Driving Car

Never lacking in innovation, Google X designed an almost zero-emission self-driving electric car technology. The company unveiled a car concept that features a model without pedals, nor steering wheel, in May 2014 at a Google convention.

Seven months after revealing the car design, the team behind the project introduced the first fully-functional prototype, proving that there's nothing too ambitious for the brains behind the company. At the beginning of 2015 and after securing permits from the state, the team began testing the car in the San Francisco Bay Area. If after several years of continuous testing, the self-driving car produces positive results, Google will begin mass-producing the said car and make it available to the public.

Included in the pre-installed equipment in the autonomous car is a $70,000 LIDAR system which uses 64 beam lasers fired from a mounted range finder to generate an extremely detailed 3D representation of the immediate environment. Combined with the pre-programmed data of the street-level maps of the world, the 3D map generated by the lasers are used as basis for the car's movements and decision-making.

During the initial testings conducted in mid-2015, most of the issues encountered by the self-driving car included safety concerns when there was snow or heavy rain. There were recorded instances when the car relied heavily on historical route data that it did not obey recently-adjusted traffic lights. When the car was faced with complicated unmapped locations, it switched to a slower and inefficient extra-safe mode. The 3D map it generated was unable to differentiate harmless light debris from heavy obstructions.

Despite all these difficulties, Google announced that as of June 2015, the electricity-powered self-driving car had traveled over a million miles. The prototype had encountered over 150 million other vehicles, more than 600,000 traffic lights and at least 200,000 stop signs. Confident that five years would make a difference, Google is gearing to release the autonomous car by 2020.

State-Wide Electromobility

The German government, in the hopes of making good with their commitment to provide country-wide electric mobility, channeled the experience and expertise of more than 30 revered German institutes, putting together industry experts for a project referred to as the "Fraunhofer System Research for Electromobility". The members are not solely from the field of engineering since the whole thrust of the project is to create a holistic interdisciplinary study that will lay down the foundation towards electromobility.

The project has already given birth to two vehicle prototypes: "AutoTram" and the "FreccO", both powered by electricity. Fraunhofer-Gesselschaft received more than 30 million euros as financial support from the German Federal Ministry of Education and Research.

The five main thrusts of the research are as follows:

Energy sourcing, power distribution, and electricity conversion: Electric vehicles are more ideal and efficient when their power is sourced from renewable resources.

Power storage technology: For more Germans to embrace electric-powered vehicles, it is crucial to invest in a mobile energy storage that can provide a steady supply of electrical energy.

Vehicle design: The shift towards electric-powered motors are revolutionary, but vehicle design must conform to modern day standards of public transport.

Socio-political integration: Introducing electromobility and technical integration of the demonstration vehicles (AutoTram – a tram system that is not installed on a rail network; and FreccO – electricity powered sports car) to the rest of Germany will require support from elected leaders of the country.

Testing and mass production: Aside from functionality, electric cars must conform with standards applied to petrol-powered cars, especially in terms of safety, comfort, and reliability.

Chapter 4:

The Cost of Electric Vehicles

Today, the average cost of an electric vehicle in the United States is $28,000. Higher quality electric cars are more expensive than this and there are electric cars available at prices lower than the average; however, it can be expected that they are the relatively low performance types of electric vehicles.

Despite various government initiatives, electric cars remain to be a small minority in terms of the global car sales. This is not enough to make a dent on the annual fuel consumption of gas, and global gas emission by the transport sector. The absence of full policy support from government leaders will only hinder the growth of the electric car industry, thereby driving technology-related costs upward.

Policy interventions are needed to increase the demand on electric cars to encourage car manufacturers to produce electric cars in larger volumes, which in turn decreases the overhead costs of manufacturing electricity-powered cars. These increases the car's viability amid a competitive automotive industry.

Over the past five years, electric cars enjoyed better efficiency compared to their petrol-powered counterparts. This drove down running costs of electric cars amounting to a measly 20 cents on the dollar cost of running vehicles powered by gasoline in Europe and the United States.

Every American using an electric-powered car has saved $2,000 over the last five years. This can be attributed to vehicle sizes and higher fuel mileage on petrol-powered car compensated by the fuel taxation break in Europe.

The most expensive electric vehicle in the market is available at a price of $105,000. It is called the Tesla Roadster and is manufactured by Tesla Motors. As you would expect, the Roadster offers great overall performance. It has a range of 244 miles, impressive acceleration times, and a 288 horsepower engine. In addition, the Tesla Roadster has a solar panel installed in its roof to help with re-charging the battery packs at no additional cost to the owner. Moreover, the Roadster's cooling system can also be powered by the solar panel.

Another popular electric vehicle of the contemporary is produced by Tesla Motors as well. This vehicle is the Model S, which costs around $50,000. This car carries the same engine qualities as the Tesla Roadster. However, the car can accommodate up to seven people. The Model S also features quick battery charging, which lasts for 45 minutes only.

The Chevy Volt is another electric vehicle that is currently available on the market. This car costs

$32,500 and has a range of 50 miles. Of course, the car's lower price reflects its lower performance, but it is quite affordable to the average consumer.

Another more affordable electric car available on the market is the Nissan Leaf. This is Nissan's venture into an all-electric car technology. The Nissan Leaf has a range of one hundred miles without needing to be re-charged. This car's battery can be fully charged after 4 to 8 hours of charging on a 220-volt charging unit set at home. It is also possible to charge the Leaf for only 26 minutes by using DC quick charging posts. However, the 26-minute charging time will only charge 80% of the car's total battery.

The high cost of energy storage is thought to be a primary reason why vehicle costs are high, and the range as to how far electric cars can go between charges are still limited. The lack of charging facilities pose another challenge to the growth of electric cars.

While most of these barriers still exist, supportive policy mechanisms, performance developments, and high efficiency have led to the steady increase of electric cars on the road. Needless to say however, the bourgeoning electric vehicle industry has paved the way for the availability of electric cars in the market. Options range from economy cars, with top speeds of 37 miles per hour, to race cars at 120-300 miles per hour.

Tax Exemptions

In an effort to encourage citizens to purchase electric-powered cars and to meet emission targets by the end of the decade, many countries have started to offer tax breaks to consumers, including:

China - Ownership or circulation taxes are not applied for electric vehicles of any kind (two-wheel or four-wheel vehicles).

Denmark - Circulation taxes are deducted from tax payables for electric vehicles weighing not more than 2 tons.

France - Electric cars used as company cars are given tax exemptions.

Germany - Electric car owners are exempted from paying circulation taxes for ten years following registration.

Japan - Electic vehicles are given automobile tax deductions and are exempted from annual tonnage tax.

Netherlands - Zero-emission vehicles, including electric cars, are exempt from having to pay road tax.

Sweden - Significant deductions are given on company car taxes if they are powered by electricity. Electric vehicles are given special exemptions from road taxes on the basis of CO_2 emissions.

United States - Some states apply regular annual tax fees while others provide exemptions.

Chapter 5:

Weighing Pros and Cons

Some experts, in and out of the electric car market, would say that the future of cars is electric. Simply put, oil is a non-renewable resource that can be depleted anytime. Hence, the belief is that people will eventually have to turn to a significantly more efficient alternative to the conventional car.

This is a prime reason why carmakers are now pursuing innovations in the electric car industry. Nonetheless, some people are still indecisive about whether to choose electric-powered cars or to stick with the traditional internal combustion engine cars. To help one decide, the following presents the advantages and disadvantages of electric powered cars:

The Downside of Electric Vehicles

Electric cars are relatively expensive.

Perhaps one of the most notable, negative characteristics of electric vehicles is their expensive price compared to their counterpart. Despite efforts to make innovations with the car features and parts that would drive the total production costs lower, electric vehicles are still generally expensive for the average consumer.

Nonetheless, some studies show that in the long-run, the use of electric cars could actually save the consumer money. Such research, however, does not matter for some people as they consider the purchase price to be just as important as future savings.

Electric cars have slower speeds and shorter driving range.

Another significant factor that is considered a major disadvantage to electric cars is their limited range. The conventional internal combustion engine car can reach 300 miles before needing to be re-fueled. Electric vehicles, on the other hand, can only reach 100 to 200 miles before the need to recharge arises.

Most consumers would prefer a rather flexible vehicle, which could handle both short distances and long travels. This was a problem considered by car companies long ago; hence, efforts on improving the mileage of the electric vehicles are continuing as strong as ever.

Electric cars also tend to be smaller and have lower top speeds than the traditional car type. The average top speed of internal combustion cars is 124 mph, whilst affordable electric vehicles usually run up to 95 mph. In addition to this, electric vehicles tend to be heavier than the conventional cars. This is primarily because of the batteries, controllers, electric motors, and chargers. Having accessories in electric vehicles such as radios and an air conditioning system could contribute to the faster drainage of the car's battery pack.

Charging electric cars may take time.

Compared to simply refilling a depleted gas tank, which often takes ten minutes tops, electric vehicle's recharging time also poses a notable disadvantage. Drained electric cars cannot be utilized for several hours until they are fully recharged.

Therefore, electric cars are not suitable for long, continuous travels. Car companies are already working on innovations aimed at making the charging process faster for electric vehicles. Also, as mentioned earlier, they are looking at the possibility of battery swapping instead of recharging.

Some electric car designs are counter productive to the vision of widespread carbon emission reductions.

Emissions also come from the electricity source utilized to charge the electric vehicle's battery. If the electricity is actually produced, the environmental benefits of electric cars are significantly decreased.

However, if the source of electricity happens to come from a renewable resource, then electric vehicles can really be the most environmentally friendly   cars.

Silence can be a disadvantage.

Electric cars reduce noise pollution, but for drivers who are used to hearing incoming vehicles, or just enjoy the raucous sound of their engine revving, may find it hard to adjust to the quiet electric car system.

Electric recharging stations are sparse.

Unlike gas stations, which are at least within a hundred miles of each other in the vastest regions of the United States, electric recharging stations are uncommon even on American soil.

Many potential consumers fear that they might find themselves stuck with their car in the middle of a long trip due to a depleted battery.

The Upside of Electric Vehicles

Manufacturing costs are on a steep decline.

Although it is more expensive to buy electric vehicles, they are actually cheaper to run than the conventional car types. Generally, it is the battery that makes an electric car expensive.

Luckily, current trends show that the future is actually looking at a price decrease for batteries. This is because of the growing number of companies manufacturing electric car batteries, forcing competitive prices in the marketplace.

Electric cars reduce noise pollution.

Another factor that makes electric cars unique is the cars' silence while in use. The most notable characteristic of electric vehicles is the minimal noise that can be heard both from the inside and out. This makes the car appealing for people who really despise the noise produced by the internal combustion engine cars.

Electric cars have lower maintenance costs.

Electric vehicles are actually mechanically simple compared to its counterpart. This makes them less costly in areas of operation and routine maintenance.

The conventional internal combustion engine cars need maintenance for the tires, wheels, fuel, warning lights, engine, lights, bodyworks, and electrical systems.

Electric cars, on the other hand, do not require as much upkeep, due to the simplicity of the mechanisms working under the car's hood.

Government provides incentives for electric car usage.

There are also incentives that are waiting for consumers who will prefer electric vehicles. Electric car owners get to pay lower taxes compared to people who own other car types. For instance, the vehicle registration tax can be reduced up to 50%. In the government's eyes, more taxes are to be paid by owners of vehicles that are highly pollutant.

Another advantage that electric vehicles have compared to conventional vehicles is the wide availability of the power needed to run the car. Power can be harnessed from any electricity source. Such sources are generally available in most households and businesses.

Mechanically-speaking, electric vehicles have better design than their petroleum-based peers.

Electric vehicles can accelerate well. There is no need to "rev up" while using an electric car to reach the maximum strength. The average acceleration of electric vehicles from 0-60 miles per hour is 4-6 seconds.

Internal combustion engine cars, however, have the average acceleration of 0-60 miles per hour in 8.4 seconds.

Electric cars are generally safer to drive.

With many of today's scientists focused on making an electric-car-dominated society, most spend their time making sure that the capabilities of electric cars outmatch that of gas powered cars.

Most of the proposed designs undergo complex testing procedures and fitness tests to make sure that the electric cars enjoy continuous supply from the battery and that all possible safety measures are in place.

Electric cars help better the environment.

Electric vehicles are seen as very good alternatives to the conventional cars primarily because of the environmental benefits they can provide. Cleaner air is possible with electric cars; this is due to the zero emission associated with them.

The use of electric cars can help reduce carbon monoxide and hydrocarbon by 98%. Consequently, electric vehicles can help reduce long-term and short-term environmental problems and that is very important to some consumers. In addition to this, batteries used in electric cars can be recycled easily. This minimizes the disposal problems that batteries usually face.

Chapter 6:

The Consumers' Perspective

Ultimately, trusting the relatively unknown future is what it takes for consumers to shift from the conventional cars to electric powered vehicles. Even today, most consumers know a pretty small amount of information about the new technology in the auto industry. Hence, the faster the information reaches the knowledge of the average consumer, the more rapidly the market may expand.

One of the major issues influencing the consumer perception is range anxiety. As mentioned earlier, electric vehicles can travel from 100-200 miles while the conventional cars top out at 300-400 miles. Although most people are concerned about this issue, it has been found in a study conducted by the Oak Ridge National

Laboratory, that such concern is not fully rational. The reason behind this is, based on the research, finding that in the United States average drivers travel no more than 35 miles a day. Such distance is indeed, within the capacity, even in the low performance electric vehicles.

The consumers' concern about the mileage is largely due to the fact that the driving population has been so used to a travel technology where range is never a significant concern. Aside from range, there are several other concerns consumers are raising when it comes to electric vehicles. Such skepticism can be largely understood considering that with electric vehicles, consumers are not just embracing a car's new feature or product, but they are also adapting to an entirely new technology, infrastructure, and lifestyle.

Indeed, there are several people who remain adamant about shifting to electric vehicles. Nonetheless, there are others who jumped into the relatively novel car type from the onset of its introduction. The early adopters were profiled as similar to those who responded positively with hybrid cars. They are young high earners, who use their electric cars as a secondary vehicle.

Most of them are situated in Southern and Northern California, where charging infrastructures are set available and the weather is good. In addition to this, the majority of early adopters have private garages with an available electric power supply. They were also found to be politically active and are highly concerned about the country's dependence to imported oil. Moreover, the early majority was found to be environmentally sensitive and willing to pay for convenience.

Based on a global study, the majority of the driving population would be willing to pay less than $30,000 for an electric vehicle purchase. However, the majority of the consumers stated that they are not willing to pay the premium. The research also revealed that consumers tended to become interested in electric vehicles due to the increasing price of fuel. Nonetheless, the majority viewed 8 hour-charging of an electric vehicle to be unacceptable for them.

Another study showed that there were three classifications of electric vehicle buyers in the United States. The first group was labeled the *advocates*. Consumers in this particular group were completely dedicated to green initiatives and they comprised of 21% of the study's population.

The second group was the *moderates*. Consumers in this group encompassed 66% of the studied population. They are the people who are committed to some green programs, at times.

The last group was called the *resisters*. Consumers belonging to this group were found to be uninterested in even considering joining any green initiatives and they made up 13% of the sample population.

In a study by the Bloomberg New Economy Fuel, it is shown that the three most important costs considered by consumers before deciding to own electric cars are: gasoline costs, maintenance costs, and battery costs.

The same study found that for consumers around the world to embrace electric vehicles, at least one of the four items below must happen:

1) Battery costs must start to go down.

2) Customers should realize that driving electric is the better car option.

3) Car manufacturers must be content with lower profit margins.

4) Government should provide incentives to drive down other related costs.

Chapter 7:

The Future

"With more research and incentives, we can break our dependence on oil with biofuels, and become the first country to have a million electric vehicles on the road by 2015"

- President Barack Obama, 2011 State of the Union

Surely, the Obama administration wants to ensure that the United States will be the leader in the growing industry of electric vehicle production. He's made it a goal for the country to reach 1 million electric vehicles on the road by the following year. This goal was seen to be a key milestone that would significantly reduce America's dependence on foreign oil.

Leading electric car companies have already set plans of production that made the said goal achievable. In addition to this, the Obama administration aimed at proposals that could help hasten the country's leadership in the electric car manufacturing industry. Such government efforts include improvements to the already existing tax credits for consumers, programs to support cities in preparation for the increasing demand for electric cars, as well as aid in research and development.

The Department of Energy has allotted $25 billion to fund the Advanced Technology Vehicles Manufacturing Loan Program. This program is aimed at helping vehicle manufacturing companies to expand, retool, and even rebuild improved facilities to make vehicles that are fuel-efficient. ATVM loans can provide, at most, 30% of the manufacturing expenses that would qualify for the DOE's requirements.

Nonetheless, manufacturers needed to undergo stringent and lengthy qualifying processes to be eligible. Consequently, companies turned hesitant to apply for the said loan. Hence, as of 2013, there is still $16.6 billion remaining in the fund for the DOE's ATVM loan program.

Aside from the loan program, export assistance is also available through the Global Automotive Team of the International Administration of the Department of Commerce. Further details regarding the said assistance can be accessed at www.export.gov.

Incentives and support services are not limited to suppliers. Demand-side inducements are also available to those looking for it. Tax credits for every purchase, tax credits at charging stations, and state financing are some of these incentives offered to consumers.

Despite having significant downsides and uncertainty, the electric vehicle industry is looking at a brighter, better future. The incentives and aids of the government motivate more car manufacturing companies to further their production.

Moreover, consumers are also given enticements to finally consider shifting from the traditional internal combustion engine car to electric vehicles. Also, research and development in the field are continuously exerting efforts to improve the capabilities of electric vehicles while trying to slowly pull the price down.

In the United Kingdom, the Committee on Climate Change turned up a report that there will be a modest reduction of carbon dioxide in the atmosphere with the gradual re-introduction of electric vehicles in many countries. With more people embracing the change, the grid eventually becomes cleaner, including the vehicles which draw the energy from it. The committee made it known that for the global carbon reduction goal to be met by 2050, there must be a widespread acceptance of electric vehicles.

In the succeeding years, new generations of electric cars will be produced. Some companies have announced their releases already. For instance, the Tesla Model X will be released in 2014. This model has the impressive acceleration from 0-60 mph in only 4.4 seconds. Also, this car will have all-wheel drive, which current models do not have. The Tesla Model X will be available on the market at around $60,000. Several other models will be available for the consumer in the very near future, as well.

The future that awaits electric vehicles is definitely bright. The significant efforts aimed at pursuing improvements and patronage for the industry will be the major driving forces towards the growth of the electric car industry as a whole.

In an article published by the New Energy Finance of Bloomberg, the sales figures of electric vehicles grew by as much as 60 percent in the global market in 2015. The same article made a bold prediction that by the year 2040, at least 7 of every 20 (35%) new cars sold will be electric cars.

Furthermore, Bloomberg predicts that the electric car revolution poses a huge threat to the oil industry, displacing as much as 2 million oil barrels each day by the year 2023. By the year 2040, 13 million oil barrels will no longer be needed to sustain the needs of the transport industry.

Navigant Research, a renowned oil and transportation industry observer, sees a strong growth by 2016. In a figure published by InsideEVs.com five months into 2016, electric cars have already posted 12% growth compared to the same period last year.

The five reasons why industry experts see a bright future of the electric car industry:

The price of batteries are falling.

Tesla and General Motors invested in research and development of batteries. Bloomberg predicts that the renewed interest in electric vehicles is driving battery costs down, making electric cars competitive by 2022.

Automakers and car experts predict that the cost of putting together a battery will go down to the target $150 per killowatt hour within the next ten   years.

Cheaper and long-range cars are slowly invading the market.

Addressing two of the most prominent concerns by people that hinder them from fully embracing electric cars bodes well for the electric car industry. Plug-in hybrid Chevrolet Volt is dubbed to be the electric car of the masses.

Later this year, Toyota will be announcing an alternate model for their flagship electric car line: Prius Plug-in.

Charging stations will be replacing gas-refilling stations.

The lack of charging stations and other similar service-providers makes the public anxious about the idea of electric cars. Due to various government programs and private sector cooperation, there will be more charging stations near transit stations, offices, apartment residences, and school campuses.

In California, the governor has already began building 1,500 charging stations to slowly reach the goal of having at least a million electric vehicles sold and on their road by 2023. Walgreens, Coca-Cola, and Google are working hand-in-hand to build several charging stations as   well.

Major players in the car industry are now designing electric cars.

Never the lack of foresight, car makers can no longer turn a blind eye on the threat posed by electric cars. Leading manufacturers now understand that the rise of electric cars are not a material for bandwagon marketing. From just two electric car models in 2010, the number has grown to 25 models today.

The number is predicted to double to 50 by 2020. By that same year, Ford is looking to have released at least 13 electric car models to account for 40 percent of their car models.

More countries are pressured to curb dependency on crude oil to cut down carbon pollution.

A research done by the Electric Power Research Institute and NRDC saw that 550 million metric tons of carbon dioxide emission can be reduced by widespread usage of electric vehicles. This is equivalent to the reduced emission of 100 million petroleum-based cars.

Almost 200 nations participated in the Paris climate accord to commit to carefully laid-out plans to reduce carbon pollution. China has surprisingly become the world's biggest electric car   market.

Conclusion

Now that you have learned about the important factors regarding electric cars, you can finally decide if you want to check one out for yourself. Plus, a little addition to your knowledge base doesn't hurt, right? It's good to know about new innovations because it keeps you in the know and up-to-date in a world where electric cars are looking to be the trend of the next decade and beyond.

At this point, you might still be skeptical about driving electric cars, but once you begin thinking things through, comparing the pros and the cons, you'll come to understand that the future is headed this way for good reason. As a responsible citizen of the world, simply taking up the wheel and driving electric cars go a long way in preserving the one and only planet we can call our home.

Hopefully you learned a thing or two in this short book. If so, share the message with others in your community! Thank you and good luck in your own journey!

Sources

Argueta, R. (2010). *A Technical Research Report:The Electric Vehicle.* Santa Barbara: University of California.

Bulk, J. v. (2009). A cost- and benefit analysis of combustion cars, electric cars and hydrogen cars in the Netherlands. *Wageningen UR*, 1-72.

Deloitte Network. (2011). *Unplugged:Electric vehicle realities versus consumer expectations.* UK: Deloitte Touche Tohmatsu Limited.

Denton, T. *Alternative Fuel, Hybrid And Electric Vehicles.* [Place of publication not identified]: Routledge, 2015. Print.

Department of Energy. (2011). *One Million Electric Vehicles By 2015.* United States: DOE.

Federal Ministry of Education and Reseach. (2015). *Electric Mobility – Rethnking the Car.* Germany: BMBF.

Hwang, R. (2016). *Future of Electric Vehiclesis Bright.* Retrieved July 7, 2016 from <https://www.nrdc.org/experts/roland-hwang/future-electric-vehicles-bright>

Kodjak, D. (2012). *Consumer Acceptance of Electric Vehicles in the US.* Washington DC: International Council on Clean Trasnportation.

https://www.iea.org/publications/freepublications/publication/Global_EV_Outlook_2016.p

Place, Project Better. (2008). Retrieved March 3, 2013, from <http://www.projectbetterplace.com/dong-energy-and-california-based-project-betterplace-

Power, J. (2011). *The Slow Road to Growing Green.* Power and Associates White Paper.

Randall, T (2016). *Here's How Electric Cars Will Cause the Next Oil Crisis.* Retrieved July 7, 2016, from <http://www.bloomberg.com/features/2016-ev-oil-crisis/>

Todd, J., Chen, J., & Clogston, F. (2013). *Creating the Clean Energy: Analysis of Ellectric Vehicle Industry.* Washington DC: International Economic Development Council.

Printed in Great Britain
by Amazon